INTELLIGENT UNIVERSE

ILYAN KEI LAVANWAY

Ilyan Kei Lavanway
CreateSpace

INTELLIGENT UNIVERSE
Copyright © 2014 Ilyan Kei Lavanway
All rights reserved.

No part of this book may be reproduced in any form whatsoever, whether by graphic, visual, electronic, film, microfilm, tape recording, or any other means, without prior written permission of the author, except in the case of brief passages embodied in reviews and articles.

The thoughts and opinions expressed in this work are the sole responsibility of the author. This work does not necessarily represent the official position of any organization.

Images included in this work are from various unknown artists and are publicly available on the internet.

ISBN-13: 978-1494905910
ISBN-10: 1494905914

Published by Ilyan Kei Lavanway at CreateSpace

Library of Congress Control Number: 2014900516

Contents

Preface	....	i
Chapter One	....	1
Chapter Two	....	3
Chapter Three	....	7
Chapter Four	....	17
Chapter Five	....	21
Chapter Six	....	25
Chapter Seven	....	31
Chapter Eight	....	35
Chapter Nine	....	39
Chapter Ten	....	47
Chapter Eleven	....	53
References	....	59
About the Author	....	61

I. K. Lavanway

Preface

The promise of God unto all who obey the Word of Wisdom: They "shall find wisdom and great treasures of knowledge, even hidden treasures..." (Doctrine and Covenants 89:19)

One fundamental question I hope and pray you will answer for yourself after reading this work: Are we mere creations of God, like the trees, the animals, and the stars, or are we literal offspring of God?

An idea came to me Wednesday morning, 19 March 2013, at around 0700 hours, shortly after I had put my son on his school bus. I do not even remember what triggered this thought, but I feel it came as a result of many questions I have been pondering for a long time. The thought came into my mind while talking to my wife about some unrelated article she had read on a news website.

I immediately felt I should make a written record of this idea. I also felt I should share it with my mother over the phone. She lives almost 3,000 miles away.

I confided this idea to my mother and she felt a strong impression that I have come upon something big, perhaps dangerously big, and that I should exercise prayerful caution in deciding when and with whom I share these thoughts.

She felt that if shared with the wrong people or at the wrong time, someone might take the idea and twist it. She could not elaborate on that feeling, but I trust her advice. I have confided to her my deepest thoughts throughout the years of my life, and never to my recollection has she told me she felt impressed that I should keep my ideas from others. Never until now, with this particular idea.

Having recently read the book titled Visions of Glory as told to John M. Pontius, I am now certain that I am not the only one who thinks along the lines of reasoning I am about to share.

I am not a man of particular import or public renown. As of the time of this writing, I am not in any position of leadership or authority to speak to the world or to the Church, in part or in whole, on any of these matters. I am one man with an opinion and an inquisitive mind and a desire to share what I learn with others.

Do not take these thoughts as doctrine. Look to the living prophet and his two counselors and the Quorum of the Twelve Apostles of The Church of Jesus Christ of Latter-day Saints for doctrinal purity and correct guidance.

Having said that, I feel it is now appropriate to share the following thoughts. Please bear in mind that this is a work in progress and is subject to future updates, corrections, and changes, as I learn more. I realize that many readers may scoff at the contents of this work. That is their prerogative. Nevertheless, I feel these thoughts are worth sharing for the sake of those who will stretch their minds heavenward, study further for themselves, and seek to know their Creator on a deeply personal level.

Intelligent Universe

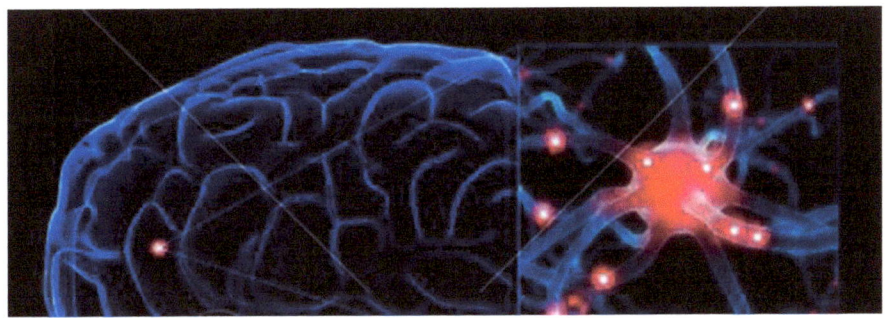

CHAPTER ONE

Have you ever wondered what the universe really looks like? Look at pictures of the human brain. That is what the universe looks like.

Stars throughout the universe generate tremendous magnetic fields. Magnetic forces within stellar atmospheres are often twisted into extreme concentrations resembling enormous tornadoes. Consider the incredible amounts of energy in solar flares and filaments.

Gravitational fields in the vicinity of stars are tightly contoured. Live stars combine matter to produce various elements. Some of the elements assembled by stellar activity are blown outward into neighboring space. How much of those elements exceed the escape velocity of the star in which they are born?

I think many of the elements produced in and released from stars are not merely blown outward by stellar winds, or by shock waves resulting from the supernova deaths of stars, but are transported to distant regions of space, far beyond the reach of stellar winds and supernova shock waves.

I believe stars generate, contain, and maintain naturally occurring wormhole terminals. I further believe that all of the stars throughout the entire cosmos are interconnected by a network of these natural wormholes.

Enoch witnessed that the earth itself is a living, intelligent, sentient being, capable of communicating with its Creator (Moses 7:48-49). It stands to reason that stars and galaxies are living organisms, individual components of a vastly greater whole.

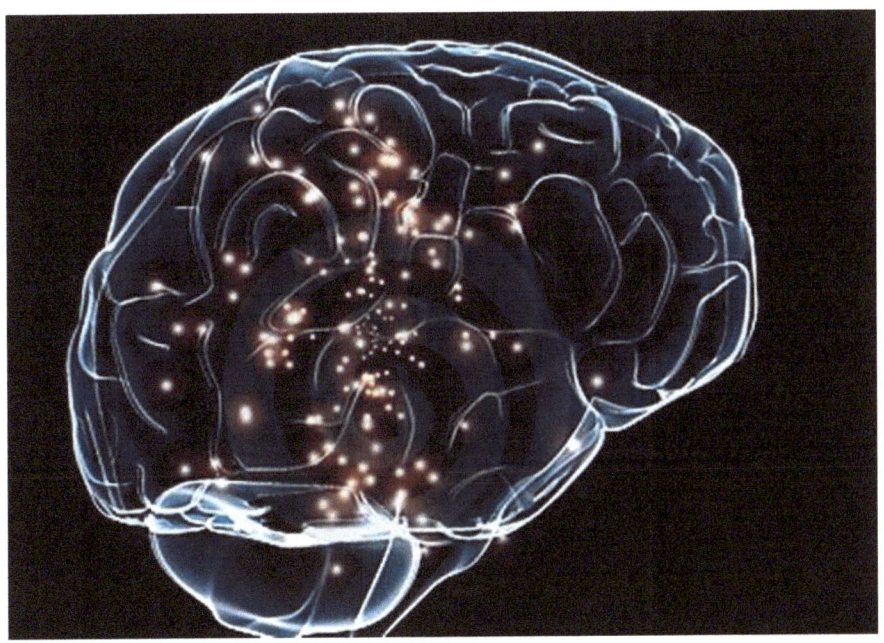

Chapter Two

The entire universe is alive and intelligent. It is designed and patterned after the human brain. It is patterned after the mind of the Creator Himself.

I believe every star is connected to at least one other star at any given moment by a wormhole whose opening is generated by the dynamic forces working within the star. Stars and their connecting wormholes function more or less like neurons in the human brain that pass electro-chemical packages to and from various regions for various purposes.

Are synaptic gaps symbolic of the veil that separates mortal man from the physical presence of God? That veil, or gap, if you will, does not impede our prayers and thoughts and feelings, and the intents of our hearts, from reaching God, in spite of the astronomical distances between our location and His. Nor does it prevent God from communicating with us and tangibly influencing our physical environment when expedient. Somehow, communication and interaction between God and mortal man is completely independent of spatiotemporal displacement.

I imagine that the wormholes terminating at each star are extremely dynamic. They may open and close at various intervals, remaining open for various seasons or durations of time. One opening in one star may shut within that star and then open in a different star. Where one closes another opens, thus there are always multiple pathways providing uninterrupted access between all points in space, and in time.

This may explain the tunnels of light described by many who have had near-death experiences, or who have actually died and then been given the opportunity to be revived to continue their mortal probation.

The universal wormhole web may also explain why resurrected beings who have appeared to mortal men are often described as descending and ascending in a pillar or shaft of light. Joseph Smith described this phenomenon associated with at least three visits he received from the resurrected prophet, Moroni in September 1823. Joseph Smith specifically used the word "conduit" when describing Moroni's departure. (Joseph Smith – History 1:43-47). The same phenomenon accompanied God the Father and His

Son, Jesus Christ, when they both personally visited Joseph Smith in a grove of trees in the early spring of 1820 (Joseph Smith – History 1:16-17).

Many scriptural accounts describe various individual prophets being caught up into exceedingly high mountains to commune with God or with angels sent by God (a few examples include: 1 Nephi 11:1, 1 Nephi 11:18-20, 2 Nephi 4:25, Moses 1:1, Revelation 21:10). How are they transported to such distant places and then returned to their localities without anyone declaring them missing for months or years?

I. K. Lavanway

Intelligent Universe

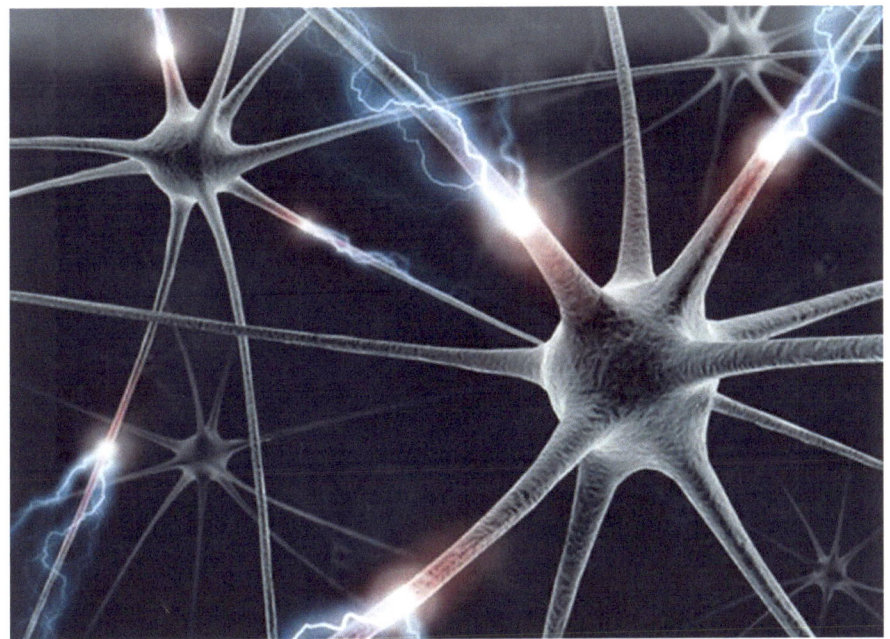

CHAPTER THREE

Travel through the wormholes would be virtually instantaneous from the perspective of anyone outside the wormholes. Think of these wormholes as freeways with various on-ramps coming from stars, and off-ramps going to stars and their neighboring regions of space. Everything is connected to everything else.

That means time travel is every bit as possible as space travel. It is all the same thing. Prophets have done it throughout history. Angels do it routinely.

Christ's atonement makes time travel possible. Does He not command time, just as He commands the elements?

How else can the Holy Ghost constantly dwell within multitudes of righteous individuals that have been baptized by proper authority and have received the Gift of the Holy Ghost? (Doctrine and Covenants 130:22) How does the Holy Ghost, who is but one individual personage, influence, warn, comfort, enlighten, and testify to the minds and hearts of multitudes of individuals around the world who have not yet received the gift of His constant companionship, but who are receptive to specific truths at various moments? (Doctrine and Covenants 130:23)

There must be a temporal loopback that, when viewed from our perspective, takes the factor of time out of the equation.

He visits you, loops back to the same time and visits me, and then loops back to the same time and visits someone else, and so on, without limit. He is constantly laboring and moving through time to minister to each of us in such a personal, efficient, and effective manner, that we all perceive His presence and feel His influence simultaneously, even though He is not actually in more than one location at a time. And then there is the warning from God, saying, "My spirit shall not always strive with man." (Doctrine and Covenants 1:33).

Visualize the use of temporal loopback another way. If you were to line up everyone in whom the Holy Ghost dwells at any given moment, the line of individuals would stretch some distance through space. Now, consider the path traveled by the Holy Ghost as He loops through time from one individual to the next, to the next, and so forth.

A looping trajectory through time, being displaced through space, creates a spatiotemporal tunnel. In other words, a wormhole. Perhaps some wormholes are created by the trajectory of the personage moving himself through time and space, like magnetic fields are generated by electrons moving through space.

A metaphor can be appended to this visualization. The displacement of the loopback through space creates a spiral trajectory that might resemble the stitching that joins tapestries together.

Are we not bound together as eternal families by the Holy Spirit of Promise when, through our personal faithfulness, the matrimonial sealing ordinance is ratified?

There are surely far more profound implications to this concept than my simplified mortal mind can conceive or express. Some implications are undoubtedly impossible and unlawful for mortal man to perceive and to utter.

As for the individual ability to move oneself through space and through time, it has been done since time began. Time is a substance given to mortals by God.

Our present paradigm is that we are subject to time's linear, forward flow. However, we can learn to utilize time like a skilled mariner uses the wind. The mariner is subject the wind, but learns to harness the wind to move himself, and the great mass over which he has stewardship, across great distances to deliver and retrieve messages, objects, and people who are prepared for the journey.

Sooner or later, some of us will learn to do this. We will be able to do it by the power of the Priesthood and by the power of faith in God, independently of any structure or machinery built by man.

True story: One evening, as I was preparing to go home after working in a Temple of The Church of Jesus Christ of Latter-day Saints, I overheard a Temple worker tell of how he and his wife had been driving on a long road trip to see family who needed them. The trip usually took twelve hours. This fellow and his wife were not driving particularly fast, and they anticipated reaching their destination at around 0300 hours. They had even called ahead to let their family know that that was when they expected to arrive. They pulled into the driveway at their destination at midnight, a full three hours earlier than expected. They could not explain it. It turned out that they needed to be there early, and somehow God made it happen.

Another true story: When I was about twelve years old, the father of my Sunday school teacher shared a personal experience about how my Sunday school teacher was almost cut to pieces by some farm machinery when he was a kid. This father, who was standing a considerable distance away, perhaps more than a hundred yards when he saw what was about to happen, somehow ran to his son and carried him to safety in a split second, far too short a time to account for the distance he had to travel on foot to get to his son and then to move him safely out of harm's way.

Personal experience: Around the year 1990, I was visiting my parents at their home in Washington State during a break from college. I was sleeping on the sofa-bed in their living room. An unusual phenomenon occurred. To this day,

I don't know what to call it, because it was not merely a dream or a vision. It was something else.

At one point, I saw a human brain, by itself. It seemed to be floating in black, open space. This brain appeared radiant. Initially, both hemispheres appeared equally luminescent. A few seconds later, the intensity of light emanating from each hemisphere began to alternately increase, first one hemisphere, and then the other. The tempo and the amplitude of this luminescent oscillation gradually increased, producing a powerful resonance, perhaps somehow capable of creating a personal wormhole or spatiotemporal conduit through which one could travel.

I sensed that this was a demonstration of the fact that the human brain has the potential to become the primary motor organ for the human body. In other words, the brain is a means of conveyance, potentially affording individuals the ability to transport themselves from one place to another, and possibly from one time to another, without dependence on any artificial contraptions or technologies, and without any interface between corporal appendages and matter.

The prospect of being able to move about solely by using my brain seemed more natural to me than walking on my legs and feet. Of course, human beings will always have legs and feet. After all, we are created in the image of God.

The experience shifted, and I was flying through the air over a wheat field near the town where I grew up. My brother, two years my junior, was standing next to a power pole, watching me.

I was not merely levitating or drifting about. I was purposefully flying with considerable velocity, and I had the ability to precisely control my orientation and trajectory at all times, simply by thoughts and feelings. There was no learning curve. I was doing it as if I had always been able to do it. Of course, at the same time, I realized that this was much more thrilling than walking or running. I was calling out to my brother to watch me as I did looping maneuvers under and over the power lines.

During the time I had this experience, I was in Washington State and my brother was stationed in San Diego, California, serving in the United States Navy. We both had busy lives, and we had not been in contact with each other for months, perhaps years.

I awoke and immediately called my brother on the telephone. If memory serves me correctly, I did not even greet him when he answered. I simply blurted out something to the effect of, "I just had a dream and you were in it."

He was ecstatic. He interrupted me and exclaimed something like, "I know. I was there. I was standing by that telephone pole watching you do loops over the wires."

Before I could give him all the details, he gave me the details. And he was not only accurate, but the perspective from which he described things indicated that he had indeed been watching me from exactly the position I had seen him.

Then, our conversation got interesting. I could only describe things between two points in time. My brother went on to describe things I had been doing before and after those points in time.

I cannot recall what he described. I do remember being astonished at the fact that what he described outside of those points in time meshed seamlessly with what I described between those points in time. And the perspective from which he described things was always consistent with where he had been in relation to me throughout the experience. How is any of this possible?

I have had at least two other similar experiences involving complete strangers whom I later met in person. In each case, I recognized the person as having been an active participant in the phenomenon.

In one case, the other person recognized me, too, when we crossed paths some time later. No words were spoken about the phenomenon, as it was far too personal to discuss. Outside of the phenomenon, we remained complete strangers to each other. During the phenomenon, however, we were not strangers, but had known each other well and for some lengthy period of time.

If the veil over the memories of our pre-mortal life were to be parted for a moment, we might understand such phenomena to be glimpses of relationships and experiences we had developed and purposed before we were born.

Consider also that Christ's atonement has eternal efficacy and infinite scope. That means it is effective retroactively and futuristically. Repentance and forgiveness were every bit as possible before Christ's atonement as after it. Prophets have expressed this fact (Mosiah 16:6).

There must have been a repentance process in the pre-mortal life just as there is in mortality. Among the two thirds who qualified themselves for mortal probation, not all were equally valiant in their pre-mortal allegiance to Christ (Bible Dictionary: War in Heaven).

When you look out over the throngs of people surrounding you in daily life, individuals you take to be strangers, realize that in the eternal perspective not a one of them is a stranger to you, nor to each other. Nor are you a stranger to any of them (Doctrine and Covenants 76:94).

The only hint of an explanation I can think of comes as a whisper to the prayerfully searching soul, and is carefully woven into LDS Hymn 97, Lead Kindly Light. I was instructed by someone I esteem as a dear friend, who is a patriarch and who was serving as a counselor in a Temple Presidency, that this particular hymn calls to mind things that are so sacred they cannot be discussed. Not even in the Temples. So, I won't say any more about it. But, you can ponder it and ask Heavenly Father to teach you what it means. I still have a lot to learn. Perhaps you will appreciate what John M. Pontius relates on pages 94-99 of his book, Visions of Glory.

Precedent for the cosmic network of wormholes is indicated in the account of the Lord explaining to Moses the dynamic and infinite scope of Creation.

Countless worlds have been and are being created (Moses 1:33). Infinitely many worlds are yet to be created. (Moses 1:37-38). They are all created for a purpose, and for the eternal benefit of the offspring of God.

We, and countless others like us elsewhere in the universe, are literally the offspring of God (Doctrine and

Covenants 76:24), endowed with the potential to become His equals if we desire to do so (Moses 1:39, Doctrine and Covenants 76:95).

Many worlds have passed away, and many now thrive. They are innumerable to mortal man, but they are counted by God. He knows them all. They are the workmanship of His hands. (Moses 1:35).

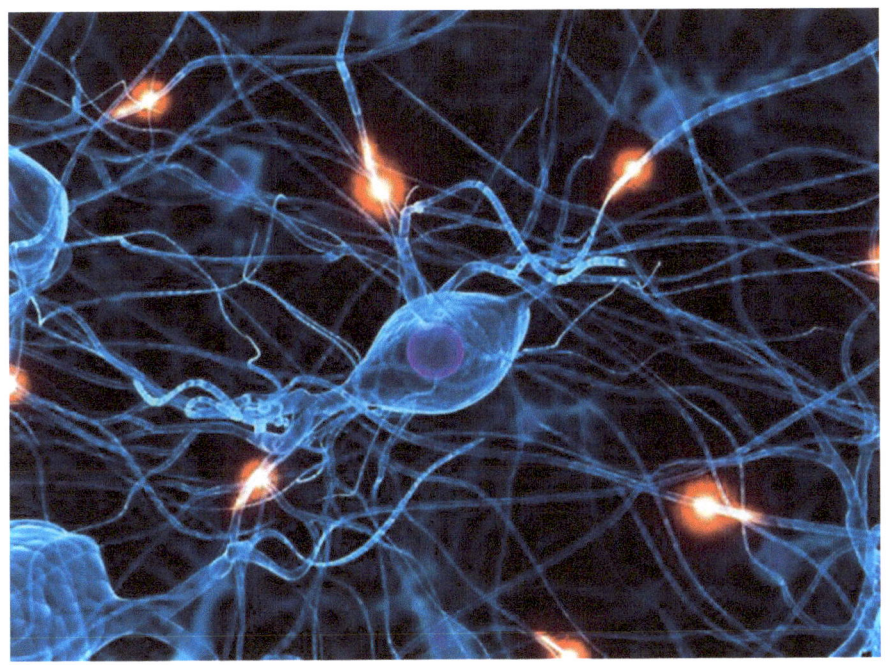

CHAPTER FOUR

Stars are born, and they live and die and are resurrected. Human beings are born, and then live and die and are resurrected. If a brain cell dies, another one takes on the tasks at hand. If a person dies, other individuals adapt and carry on. Life is divinely patterned for eternal continuity.

Likewise, when one star dies, or when one terminal closes, the pathways connected to it are rerouted to compensate for the change. Nothing is permanently lost. Functionality is preserved. Factoring in the process of resurrection opens the mind to an incomprehensibly vast and eternal splendor.

Perhaps the interstellar system of wormholes is one of the means through which matter and the elements are distributed throughout the universe. Maybe this has something to do with the so called dark matter and dark energy (dark meaning undetectable by conventional methods) that supposedly make up the bulk of our universe, but which few, if any, on Earth have been able to directly observe. Such matter may also be spirit matter that is undetectable to mortals (Doctrine and Covenants 131:7-8).

I think such an interstellar connectivity network extends among galaxies and galaxy clusters as well as among the stars of any given galaxy. In other words, there is an infinite and eternal cosmic network that fills the whole of space (Doctrine and Covenants 88:12-13), such that every volume of space pertains to and is governed by some kingdom, and every kingdom has prepared for its inhabitants some region or regions of space (Doctrine and Covenants 88:37-38).

Perhaps through this universal system of spatiotemporal tunnels, our earth was transported, fell, from the presence of God, our Heavenly Father, when it entered its Telestial existence after Adam and Eve had eaten the forbidden fruit in the Garden of Eden. That event was not called the Fall of Man for no reason.

It may be that through this universal wormhole web, the earth will eventually be removed out of its place (Isaiah 24:20) and brought back into the presence of that God who designed it and assigned it to us as the place of our mortal probation and the seat of the atonement of His Son, Jesus Christ.

I. K. Lavanway

Intelligent Universe

CHAPTER FIVE

If our eyes were capable of seeing everything in space, we would look out and see an intricate and dynamic network of interstellar, intergalactic, even inter-universal connectivity similar to the web of nerve cells that make the human brain what it is.

This interstellar network of wormholes may have been exactly what the ancient inhabitants of Babel were trying to access by building a tower to reach heaven (Genesis 11:4).

The Tower of Babel was a pyramid, or more precisely a ziggurat. Many pyramids that dot the earth have internal shafts that align with specific stars, including stars in the constellation Orion.

Dr. David B. Cohen elaborates on the significance of Orion is in his book, The Adamic Language and Calendar: The True Bible Code. The shape of the constellation Orion forms a character in the Adamic (ancient Hebrew) language that represents the name of Jehovah, who is Jesus Christ.

Aleph Ori, more commonly known as Betelgeuse, is the red giant star occupying the right shoulder of Orion when viewed as if Orion were a figure facing our earth. Aleph Ori means "my number one light" in ancient Hebrew.

Christ is "The Light" and "The Only Begotten of The Father" and "The Firstborn" or "Firstling" of the Father. In other words, Jesus Christ is the Father's number one light.

Why did the ancient Hebrews name the red giant star on the right shoulder of Orion after Jesus Christ? Why did Heavenly Father place stars in a precise pattern that, when viewed from the location in space occupied by this earth in its fallen, Telestial state, looks like the character signifying Jehovah in the Adamic language?

When Christ returns to Earth in His full glory, He will be arrayed in a red robe (Isaiah 63:2), symbolic of having tread the wine press alone, having shed His blood for all mankind. Aleph Ori is a red star. Ancient pyramids have tunnels within them that precisely align with the constellation Orion. This cannot be mere coincidence or happenstance.

Could the stars in Orion, specifically Aleph Ori, be an access to the wormhole path that leads back to the governing star called Kolob, and to the place of this earth's creation, and from there to the place of our spirit nativity in the presence of God, our Heavenly Father, regions that are not known to mortal man? (Doctrine and Covenants 133:46-50).

No one can return to the presence of the Father except by and through the grace of Jesus Christ. Some are foolish enough to think they can bypass the laws and ordinances of salvation and exaltation, supposing they can devise a clever way to circumvent the plan of God and access His power and glory while shirking individual responsibility and escaping personal accountability that comes with Celestial power (Doctrine and Covenants 121:34-40). There is no back door entrance to the throne of God.

I believe that the ancient inhabitants of Babel were trying to access the heavens by building an infrastructure on Earth that would allow them to travel through wormholes to and from anywhere. I believe those who were building the tower of Babel had the records and the knowledge of the patriarchs, passed down from Adam to Enoch to Noah.

Noah had surely preserved records aboard the ark, surviving the global flood that had occurred just over one hundred years prior to the tower project. In fact, Noah was still alive during the days of Babel. He lived another two and a half centuries after construction on the tower had begun, and about fifty-eight years after the birth of Abraham (Genesis 9:28-29, Genesis 11:10-32).

The nature of society in the days of Babel was not one of continual violence as had been the case in Antediluvian history, that is to say, preceding the global flood.

In the days of Babel, awareness of the global flood was still relatively fresh in the minds of the children and grandchildren of Noah. A group of Noah's posterity, perhaps including some of his grandchildren and great grandchildren, founded the City of Babel and set about building the tower. They were well aware of the power of God to unleash an extinction level event on the planet. Their problem was not violence, but intellectual pride and spiritual overzealousness.

Intelligent Universe

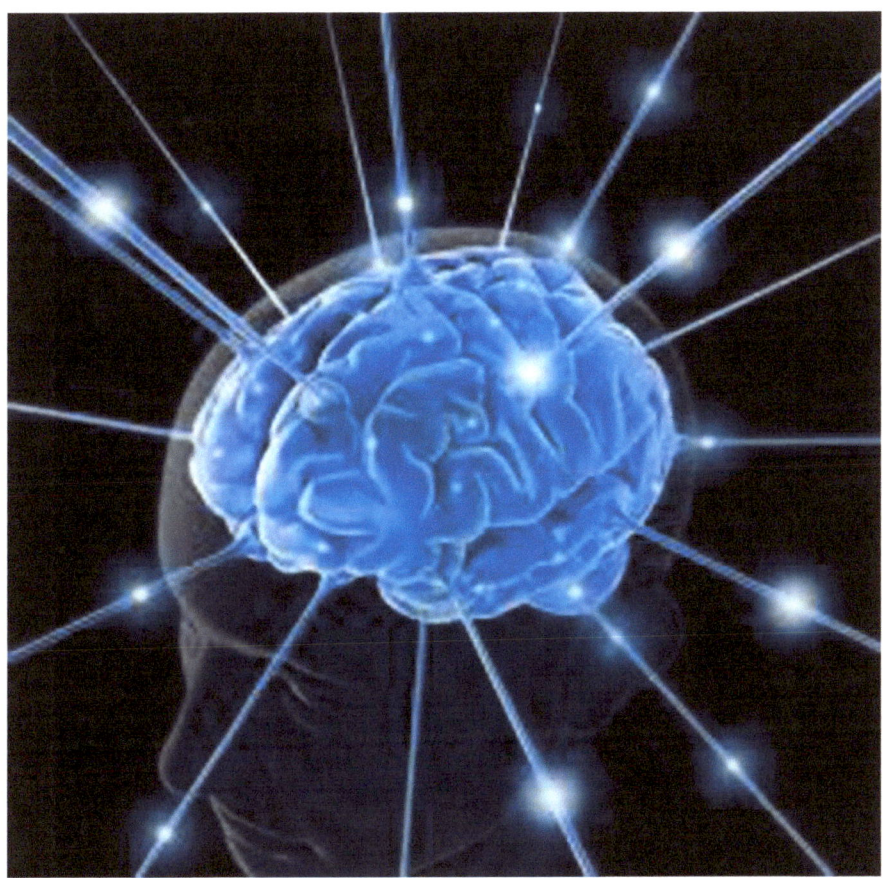

CHAPTER SIX

I feel that the builders of the Tower of Babel had at least one of the following four objectives in mind, and that they failed because of the impurity of their intent. I do not think such endeavors necessarily fail among the inhabitants of other earths where populations are strictly obedient to God. As for our people of Babel:

One: They were afraid of being driven out of the comfort and familiarity of the region where they had settled in the plains of Shinar. They knew God had commanded them to go forth and populate the earth, but that would have involved difficult and arduous travels and labors in unfamiliar regions of the earth. They probably identified with what Adam and Eve must have felt upon facing the prospect of being driven out of the Garden of Eden into an unfamiliar and changed world. The builders of the tower likely sought a way to leave the uncertainties and hardships of Earth and find a more paradisiacal environment.

Two: they were afraid of another extinction level event. The groanings of continental separation had probably begun around that time, as the Pangaean land mass was broken up in the days of Peleg (Genesis 10:25). Peleg was born 101 years after the flood (Genesis 11:10-16), which means his birth was around the time construction began on the Tower of Babel. The inhabitants of Babel knew that the City of Enoch had been translated, that is to say, quickened from a Telestial state to either a Terrestrial or a Celestial state and removed from the earth (JST Genesis 14:34, Moses 7:19-21) by God some 603 years prior to the global flood (Moses 6:10-25, Moses 8:5-13, Doctrine and Covenants 107:48-49, Genesis 7:11, Genesis 8:13). The idea of going off planet to escape another catastrophic global change was a tempting prospect. The inhabitants of Babel may have been trying to transport themselves to the City of Enoch. Their forefathers were among the people of Enoch.

Three: They wanted to become gods, or be esteemed as gods. They knew God had given Adam dominion over the earth and had instructed Adam to subdue the earth. They may have twisted that instruction to suit their own desires and not the will of God. The idea of taking control of an entire planet was not a foreign concept to them. They also knew that the divine potential and purpose for which mankind exists is to become like God, in whose image man is organized. The inhabitants of Babel lived in the recent Postdiluvian environment. They probably wanted somebody over whom to wield their substantial knowledge and power. They may have been seeking a way to access other inhabited worlds with the intent to subdue those worlds, and perhaps to bring people from distant worlds to Earth to be subjugated here. If such would have been the case, the inhabitants of Babel would have indulged themselves in an abundantly replenishable workforce, a fresh harem continually at their disposal, and the glories of lordship perpetually feeding their pride and lusts.

Four: They may have been trying to travel back to the place of their spirit nativity, the place where God dwells, in the vicinity of Kolob (Abraham 3:2-4). Perhaps they were trying to literally worm their way through the veil that separates us mortals from that God who begat our spirits. They understood the fact that every mortal human being has the divine potential to become a god or a goddess, equal to God Himself (Doctrine and Covenants 76:95). These people wanted to bypass the divine plan that leads the faithful to eternal life and Celestial exaltation. They wanted to do it their way, and take the credit for having circumvented the plan of God. They wanted to make a name for themselves among all the children of God (Genesis 11:4).

Whatever their objectives, the denizens of Babel were definitely onto something. The were about to tamper with universal access protocols that were of such tremendous import that Jesus Christ Himself came down to personally assess the situation (Genesis 11:5).

Christ had observed that because of the power of their language, which was the Adamic language (Genesis 11:1, Moses 6:5-6), which is also the language of God (Moses 6:46), and because they had begun to do this, specifically referring to the construction of a pyramid with the intentions they harbored and the knowledge they possessed, nothing could be restrained from them which they had imagined to do (Genesis 11:6).

The ancient patriarchs from Adam to Enoch to Noah to Abraham all had tremendous understanding of pre-mortal history, the Creation, the planets, and the stars (Abraham 1:31).

Around the time of Babel, approximately 2242 BC, which was about one century after the global flood of Noah's day receded, Christ confounded the language of Earth's inhabitants (Genesis 11:7). The people of Babel were subsequently dispersed into diverse groups and migrated to different parts of the earth (Genesis 11:8-9).

Many of the resulting civilizations built pyramids. There seems to have been an obsession with building pyramids among early civilizations on Earth, particularly among early Postdiluvian cultures. Why?

Unexplained mysteries continue to baffle our so called enlightened, modern civilization regarding ancient pyramids. Or that is what someone wants you to think. There appears to be some conspiracy at work covering up the historical and scientific significance and the practical applications of the ancient pyramids. But, that is a subject for another book.

I. K. Lavanway

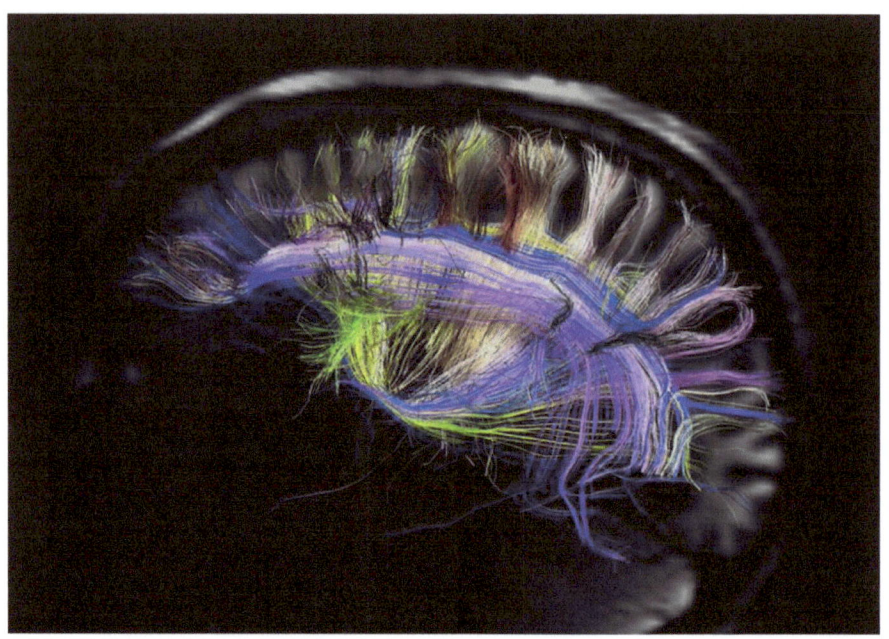

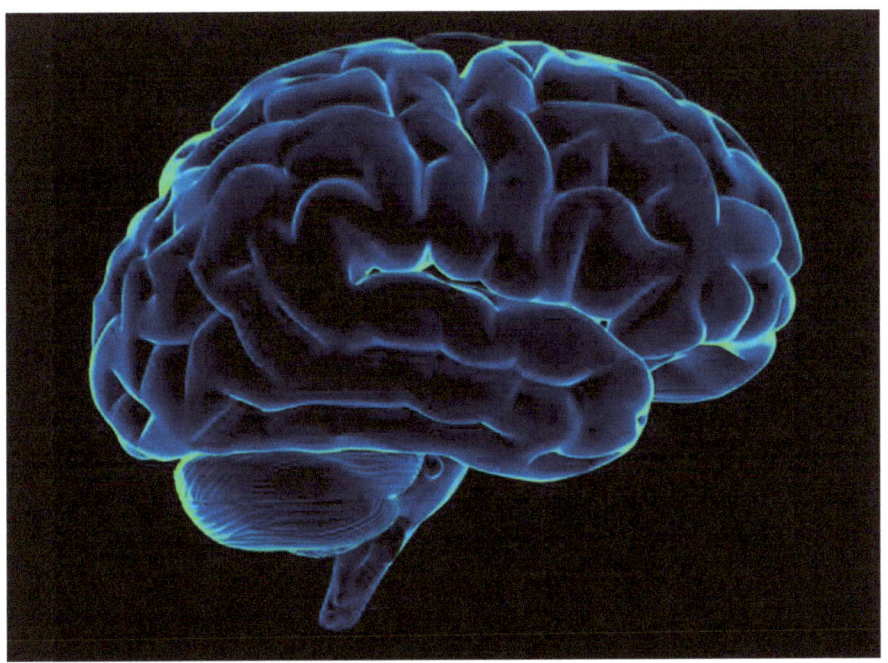

CHAPTER SEVEN

I feel impressed that there is far more truth to this whole idea than I am yet aware. I do not comprehend the ramifications of making others aware of the fact that I entertain this idea that the entire universe is modeled after the human brain, which is the express image of the mind of God, and has practical applications for the accessing of the heavens and the transporting of people and materials independently of man-made vehicles and earthly economies.

I know that we are created in the image of our Heavenly Parents, male and female. I know that God, who is our

Heavenly Father, has designed and created countless worlds, including numberless worlds inhabited by His begotten sons and daughters like us.

If we could number the particles of this earth and millions of other earths like ours, it would not begin to approach the number of inhabited worlds God has created through His Firstborn Son, His Only Begotten in the flesh, even Jesus Christ (Moses 7:30).

I. K. Lavanway

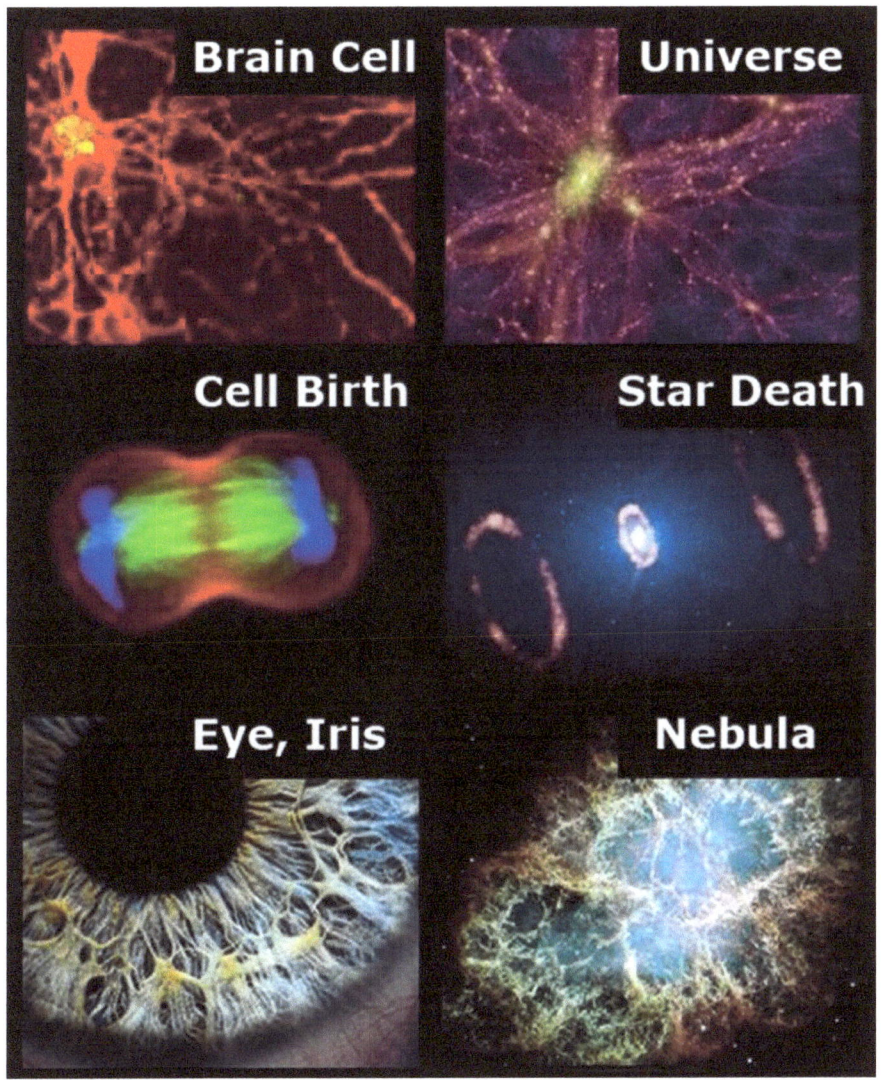

I. K. Lavanway

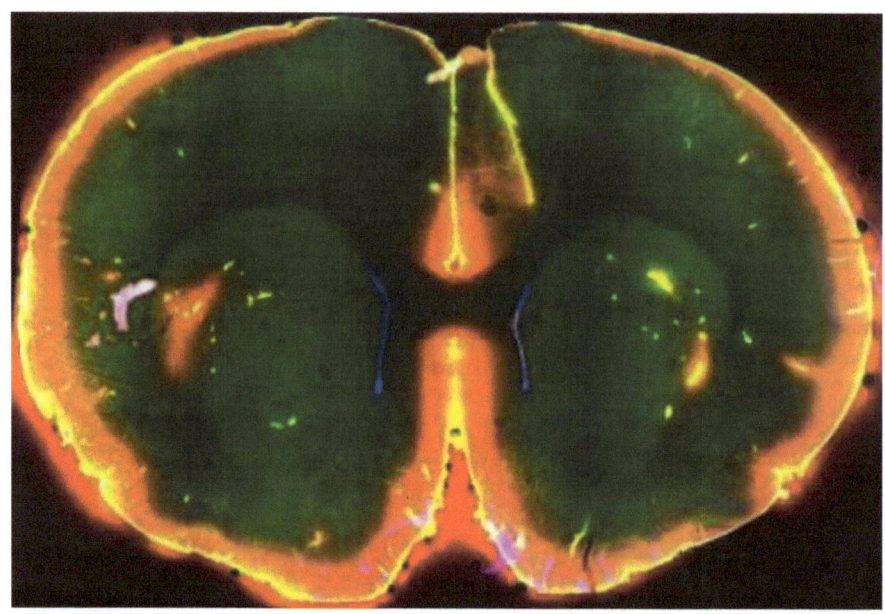

CHAPTER EIGHT

Here are a few additional thoughts. As is the case for all content in this book, these thoughts are not necessarily doctrine. They are my attempts to learn more, and I have a long way to go.

Outer darkness is a kingdom, but not a degree of glory. There is a way in for those who are ordained to such condemnation, but there is no way out for them. I imagine there must be some type of wormhole connections or portals into kingdoms of outer darkness.

Perhaps such connections function as one-way portals for the inhabitants of kingdoms of outer darkness, and two-way portals for visitors from higher kingdoms who, for whatever rare purpose, may come and go, or permit a brief and limited exposure to such an environment and then straightway shut, as in the case of Joseph Smith and Sidney Rigdon (Doctrine and Covenants 76:45-48).

It might be worth mentioning that in the case of this earth, Lucifer and his one third are still on the earth. They are not yet in a kingdom of outer darkness. They were cast out of the presence of God and confined to the earth (Revelation 12:9, Revelation 12:13, JST Revelation 12). Here, they must remain while the earth continues its temporal existence. They are, however, condemned to inherit a kingdom of outer darkness after the final judgment, following the return of Christ to this earth and following the completion of Christ's millennial reign and the final battle upon this earth (Doctrine and Covenants 88:108-114, Revelation 21:1-4).

By the time this earth, which is Christ's footstool, is resurrected and receives its Celestial glory (Doctrine and Covenants 38:17, Doctrine and Covenants 130:9), Lucifer and his one third, and the sons of perdition that pertained to this earth, will be consigned to a kingdom of outer darkness for eternity. From there, they cannot escape (Doctrine and Covenants 29:28-29).

I believe there are spatiotemporal fibers of light, and there are similar fibers of darkness. I had an experience a few years ago that gives me reason to think of such things. I included an account of that experience in my book, Earth Sink, bottom of page 133 through page 136.

If wormhole terminals are used to deposit sons of perdition into outer darkness, such terminals might have some relation to spatiotemporal fibers of darkness so as to not impart light into a place which, by divine design, must remain devoid of light. Such fibers may have an eternally confining and restrictive effect, literal chains of hell. Such an effect is the express opposite of the liberating effect of intelligence. Intelligence is light.

Just as there are stars that give off intense light, there may be stars that emanate utter blackness. These stars, if they exist, are not necessarily black holes, but something entirely different. Black holes ingest light. Light goes in. Where it goes from there, I don't know. Perhaps it stays locked in a trajectory below an event horizon, or perhaps it makes its way into another universe altogether.

Whatever the case, the fact that light can enter a black hole suggests that black holes have little or nothing to do with outer darkness. They may, however, have something to do with kingdoms that are neither outer darkness nor degrees of glory. Such kingdoms must exist for those who go away to their own places and do not merit a degree of glory but are not sons of perdition (Doctrine and Covenants 88:24,

Doctrine and Covenants 88:32) Such persons simply want to live an eternity devoid of anything to do with God or the Godhead.

CHAPTER NINE

Allow me to digress for a moment. The notion of black holes being regions of lesser brilliance might be exactly opposite of what is really going on. Black holes only look black from outside an event horizon.

Inside a black hole, everything outside is observable. Outside a black hole, nothing inside is observable.

The mysteries of God would be esteemed as foolishness by iniquitous men, and thus they are hidden from such men (1 Corinthians 2:14). Could the throne of God be at the center of a super massive black hole, or something to which a black hole is analogous?

If such were the case, all things of a Celestial order would be veiled behind an event horizon. They would be undetectable by anyone inhabiting regions more distant than the event horizon.

Terrestrial kingdoms would occupy regions of space concentric to and outside the Celestial event horizon.

Telestial kingdoms would occupy regions of space concentric to and outside the Terrestrial event horizon.

Kingdoms that are not degrees of glory would occupy regions of space concentric to and outside the Telestial event horizon.

Finally, kingdoms of outer darkness would occupy regions of space concentric to and infinitely distant from regions pertaining to kingdoms devoid of glory that are not kingdoms of perdition.

Visualize this simplistically, as the concentric layers of an onion, but in a mirror reversal of the description presented in the book Earth Sink, pages 20-21.

Here, we can think of the center as being the Celestial throne of God. The successively lesser kingdoms are positioned at further and further distances away, like the outer layers of an onion. Note that more than one event horizon is involved.

Terrestrial beings cannot perceive or access Celestial realms, because the floor, for lack of a better word, of the Terrestrial is an event horizon, as viewed from the perspective of Terrestrial beings. The event horizon perceived by Terrestrial beings veils all things Celestial.

Telestial beings cannot perceive or access Terrestrial realms, because the floor of the Telestial is an event horizon, as viewed from the perspective of Telestial beings. The event horizon perceived by Telestial beings veils all things Terrestrial.

Interestingly, Telestial beings would have no perception or awareness of the existence of the Celestial event horizon at all. They would only perceive the Terrestrial event horizon as the limiting boundary of their kingdom. Anything beyond that would not even come to mind (Isaiah 65:17).

Inhabitants of kingdoms devoid of glory that are not kingdoms of perdition cannot perceive or access Telestial realms, because the floor of their domain is an event horizon, as viewed from their perspective. The event horizon perceived by beings devoid of glory that are not sons of perdition acts as a veil concealing all things Telestial. The inhabitants of kingdoms devoid of glory would have no perception or awareness of the existence of a Terrestrial event horizon or a Celestial event horizon.

Finally, beings consigned to kingdoms of outer darkness would have no awareness or perception of the existence of any degree of glory whatsoever. The event horizon they perceive would function as a veil hiding all things of any order other than outer darkness.

The eternal increase of our Heavenly Father's dominions would necessitate an ever expanding volume of space. His Celestial Kingdom, being ever expanding and always central to all of his works, would require that lesser kingdoms oriented concentrically and external to His Celestial Kingdom be continually pushed further and further away from the center place of His Celestial Kingdom. Could this have something to do with the fact that all observable galaxies and galaxy clusters are accelerating away from each other?

Well known LDS (Mormon) scholar, Lynn M. Hilton, PhD, wrote a book in 2005 titled The Kolob Theorem. In that work, Dr. Hilton subscribes to the notion that the center of every galaxy is the Celestial throne of a god.

Dr. Hilton posits the idea that our Heavenly Father resides at the center of our galaxy, and that the lesser kingdoms occupy the bands of stars progressively more distant from the galactic center. He further suggests that outer darkness is located in intergalactic space.

I respectfully but strongly disagree. My disagreement is not so much with the construct but with the scope of his theorem.

First of all, intergalactic space is anything but dark. It is teaming with energy. More importantly, the scriptures state very clearly that the creations of our Heavenly Father, our God alone, exceed the understanding of man and cannot be numbered by man (Moses 1:33, Moses 1:37, Abraham 3:12).

Moses was permitted to see every particle of our earth. In fact, there was not a particle of it that he did not observe (Moses 1:27). That means Moses, a mortal man, was permitted to number the particles of our earth. That number, whatever it is, represents a finite quantity.

Centuries prior to Moses, Enoch understood that the particles of millions earths like ours would still be a finite quantity and would not begin to compare to the number of worlds, galaxies, and universes created by our Heavenly Father (Moses 7:29-30).

We can virtually count all the stars in our galaxy. In the infinite expanse of our Heavenly Father's creations, our galaxy, even our entire universe, is an infinitesimal speck.

Even with our mortal eyes and telestial instruments, we can see that there are countless galaxies and clusters of galaxies merely within the tiny bubble of perception we call our observable universe. There is so much more beyond what is visible to us. So much more, even within the universe we inhabit.

I am convinced that all of what we can observe in the cosmos, and infinitely more which we cannot observe, are the workmanship of our Heavenly Father, our God. To suppose that the creations of our Heavenly Father, who is God of gods (3 Nephi 4:32, Doctrine and Covenants 76:112, Doctrine and Covenants 121:32), is limited to one galaxy, is myopic at best. No disrespect intended to Dr. Hilton. These are simply my personal thoughts. Perhaps later editions of Dr. Hilton's Theorem address this.

Bear in mind that our galaxy and every other galaxy observable to us are presently in a mortal or temporal state of existence. If we can observe them with our mortal senses, they are mortal. Galaxies as we know them do not last forever. The place where God resides is a place of resurrection and eternal glory. It will never pass away or be disrupted. It is central to all His creations (Doctrine and Covenants 88:6-13).

The governing stars, the greatest of which is called Kolob, clustered around the place of our Heavenly Father's throne, are eternal, immortal, Celestial. Each governing star exercises influence over many planets pertaining to a given order, and there are many different orders of planets (Abraham 3:1-4, Abraham 3:9).

It is possible, perhaps likely, that the core of every galaxy has a throne belonging to our Heavenly Father. That would make far more sense than supposing His Celestial Kingdom is limited to the central core of one galaxy.

Our Heavenly Father has infinitely many thrones. Knowledge of that fact has been clearly restored in this dispensation. So, rather than speak of "the" throne of our God, we might do well to specify which one of His thrones we are discussing.

The only way Dr. Hilton's theorem would make any sense to me is if, in the context of Abraham's reference to Kolob and the throne of God, Abraham was referring to one of infinitely many Kolobs and one of our God's infinitely many thrones, the throne around which our Kolob orbits, the throne specifically located at the center of our galaxy.

Abraham was speaking of our Kolob, the Kolob that governs the order of planets of which our earth is a member. That is an order of inhabited, earthlike planets within our galaxy, our galaxy being merely one of infinitely many galaxies created by and belonging to our God.

If we acknowledge that our Heavenly Father may have a throne at the center of every galaxy he has created, and that he has clustered a set of governing stars including a Kolob around each of His thrones governing earths like ours which belong to a specific order created within and assigned to each of those galaxies, then we might be onto something sensible and more in line with scriptural accounts.

In our Heavenly Father's house there are many mansions (John 14:2, Doctrine and Covenants 98:18, Doctrine and Covenants 81:6). Supposing the word "mansions" is synonymous with thrones in certain contexts, it suggests an infinite number of galaxies, each having a throne belonging to our Heavenly Father at the galactic core.

Our Heavenly Father has created and continues to create countless, infinitely vast universes, the number and scope of which defy mortal man's ability to even begin to imagine, let alone observe empirically.

Other Heavenly Fathers have created and continue to create countless, infinitely vast universes as well. Proprietary divine workmanship far exceeds the scope of any singular universe, and certainly surpasses the span of any one galaxy.

Now, back to where I was before I digressed.

A close friend of mine shared with me an interesting thought that may answer a question I have had for many years. His thought was that perhaps God taps the kingdoms of perdition to seed opposition on other earths. I have a couple of lines of reasoning regarding other satans on other earths.

Chapter Ten

My first line of reasoning is this: As far as I know, among all of our Heavenly Father's spirit offspring, regardless the worlds to which they are assigned for their mortal probation, and regardless of how many wives our Heavenly Father has, there is only one Jesus Christ. Period.

Jesus Christ is the very first spirit child begotten by our Heavenly Father, hence the term firstling or firstborn. That is to say, when our Heavenly Father entered his exaltation and became a god, his very first spirit child was Jesus Christ.

Perhaps his second spirit offspring was the Holy Ghost. Perhaps his third was Lucifer. That would mean, among the first three spirit siblings, two thirds are part of the Godhead and one third is in opposition. Symbolism runs deep.

I think "son of the morning" means a spirit child born to newly exalted heavenly parents, and makes reference to the morning of their resurrection.

In this line of thinking, there would have to have been countless worlds created by Jesus Christ and populated with our Heavenly Father's spirit children long before the creation of our earth began, and long before Lucifer rebelled. That would mean the opposition encountered on worlds that pre-date Lucifer's rebellion must have come from a source other than Lucifer.

Supposing our Heavenly Father organizes his spirit children into groups assigned to specific earths, and supposing the spirits assigned to a given planet are born of one of Heavenly Father's wives, it is reasonable to suppose that at least one of the spirits in the human family assigned to any given earth chooses to rebel and refuses to keep his first estate. Likely, the one rebelling would convince about a third of his siblings assigned to that earth to rebel with him. In other words, perhaps other once righteous spirit children rebelled among each and every batch of spirits foreordained to a mortal probation on the world to which they were assigned.

While opposition may be introduced to every inhabited world by fallen spirits who pertain to that world, salvation and exaltation come solely by and through the one and only Jesus Christ.

Jesus Christ was held in reserve for a time when our earth would be created (Doctrine and Covenants 121:32). Our Heavenly Father directed the commencement of this earth's creation to coincide with Lucifer's rebellion. Our Heavenly Father had foreknowledge of that rebellion long before Lucifer was even begotten as a spirit child.

That means the inhabitants of countless worlds were waiting eons for this earth to be created and for the meridian this earth's mortal existence wherein the atonement would be completed and resurrection would be made available to every spirit child of our Heavenly Father. That is to say, every spirit that had kept his or her first estate.

Perhaps some non-linear attributes of time were in effect for the righteous inhabitants of worlds that pre-dated our earth and Christ's atonement. From our perspective, it may seem that they were required to wait a disproportionately prolonged period of time to enjoy resurrection. However, our perception of time is not necessarily their perception of time.

Einstein's theory of relativity applies to gravitational concentrations just as it does to the speed of light. Mass is a function of energy, and gravity is a function of mass. So, gravity is a function of energy. There is tremendous energy, energy being one aspect of glory, at the throne of God. That means there must be a tremendous gravitational force surrounding the throne of God. Gravity affects time.

While Einstein's theories are surely incomplete in the eternal sense, they may at least serve as a metaphor for the following phenomenon: The closer one is positioned to the throne of God, the faster time seems to flow when looking outward, and the slower time seems to flow as seen by an observer far away looking inward. Hence, at the throne of God, time moves infinitely fast, and all things, past, present, and future, are continually before God.

For someone far removed from the throne of God, looking in, time appears to stop, and nothing ever changes. Could this, in part, be what is meant when we refer to God as an unchanging God?

Perhaps worlds that pre-date our earth were mercifully positioned nearer to the throne of God, even in their fallen, Telestial estate. It is written that one Telestial body differs in glory from another, as one star differs in brightness from another (Doctrine and Covenants 76:98).

Given the fact that the inhabitants of other worlds belonging to our Heavenly Father are not nearly as wicked as the wicked who dwell on this earth (Moses 7:36), those worlds may have been permitted a less distant fall, enabling their time to flow more rapidly than time as we perceive it.

The worst of their wicked are not nearly as iniquitous as those on our earth who crucified Christ. Had Christ been born on any other world, the inhabitants of that world would not have crucified him, and the atonement would not have happened. The ancient American prophet, Jacob, younger brother of Nephi, alluded to this when he explained that even on this earth Christ would go among the Jews to be crucified, because no other nation was wicked enough to crucify their savior (2 Nephi 10:3-5).

Here, my friend's thoughts give me more to consider. Perhaps there are cases where no spirit assigned to a given world rebels, and Heavenly Father decides to introduce a catalyst by which to accelerate His plan on that particular world.

He could unleash a carefully measured portion of the inhabitants of a kingdom of outer darkness onto that world just long enough to produce the necessary effects. Those temporarily unleashed devils could be sent back to their kingdom of outer darkness before they would utterly destroy the work of God on that world. Then, should the work of God on that world falter beyond redemption, the responsibility would rest upon the shoulders of the disobedient inhabitants of that world.

The complete wasting of entire worlds may occasionally occur. Christ alludes to exactly such a possibility when he warns that if we, on this earth, were to fail to do the vicarious work of redemption in the Temples on behalf of our dead, this earth would be cursed and utterly wasted at His Second

Coming (Doctrine and Covenants 2:3, Doctrine and Covenants 128:18, Doctrine and Covenants 138:47-48).

    I wonder if Mercury, Venus, Mars, and the asteroid belt are examples of such cases, placed near our earth as reminders for a time when our earth is in the final moments of its temporal existence, and the inhabitants of this earth begin to forget, yet again, and neglect the imperative nature of vicarious work for the dead. The millennial reign of Christ is followed by a battle the likes of which this earth has never known. Undoubtedly, one of the key adversarial objectives in that final battle is the interruption of the finishing of the vicarious work in the Temples.

Intelligent Universe

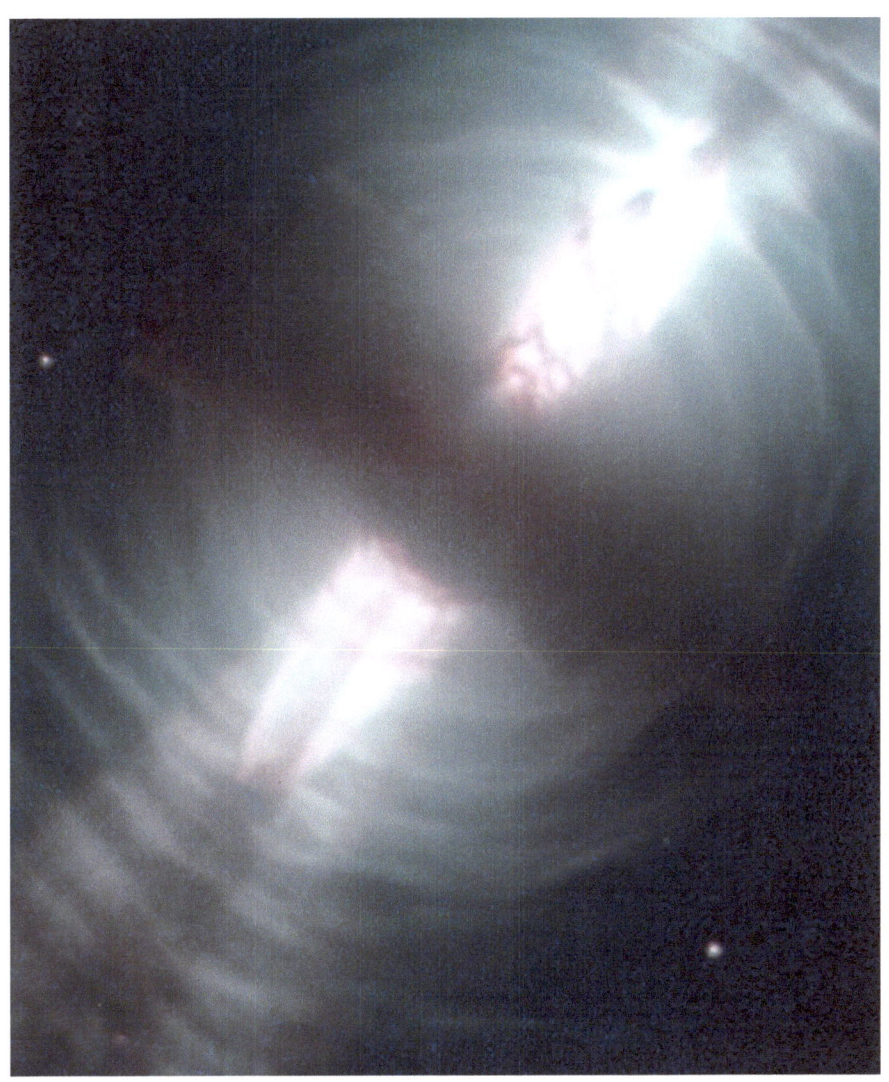

CHAPTER ELEVEN

My second line of reasoning is this: Every world needs opposition (2 Nephi 2:11-12), but not every opposition comes from a satan.

Opposition is an integral part of Heavenly Father's plan, independent of Lucifer (2 Nephi 2:15-16). The conflicting commandments given to Adam and Eve presented an inherent opposition that required them to eventually make a choice one way or the other, all on their own, with no external influence.

Lucifer unwittingly became a convenient catalyst for the acceleration, or hastening, of the plan of salvation on this earth. The timing of this earth's creation suggests Heavenly Father's foreknowledge and use of the timing of Lucifer's rebellion. In other words, not only did Heavenly Father know eons in advance that He would have a spirit son who would eventually rebel, He knew exactly when and how that rebellion would come to pass, and He planned to capitalize on it according to His divine wisdom.

This earth being the seat of the atonement of Jesus Christ suggests the need for a hastening of Heavenly Father's plan on this earth, so as not to expose it and its inhabitants to more than they can bear (1 Corinthians 10:13). If this earth should be afflicted for prolonged eons by the unrelenting and ubiquitous abominations that far exceed the evils inflicted upon other earths, even the very elect inhabitants of this earth might be deceived and spiritually perish, forever losing their Celestial inheritance (Matthew 24:22, Matthew 24:24, Joseph Smith - Matthew 1:22).

Lucifer thought he was thwarting Heavenly Father's plan by tempting Eve to eat the forbidden fruit. In reality, he was hastening the plan by orders of magnitude. What might have taken trillions of years or more happened in less than one thousand years, and Lucifer bought himself and express ticket to outer darkness. He could have tarried on the earth in its Terrestrial estate for eons had he simply left Adam and Eve to their own cognizance. Talk about shooting himself in the foot.

I suppose it might be possible for some worlds, perhaps most worlds, to go through their mortal probation with no satan and no demons indigenous to those worlds. In such cases, opposition would come by the respective Adams and Eves on those worlds eventually asking Heavenly Father, or contemplating for themselves, how it is possible to obey both the commandment to multiply and the warning to not eat the fruit of the tree of knowledge of good and evil.

Being intelligent, they would eventually realize that they must make a choice. Sooner or later, they would have to observe that the only way for them to multiply and enable the plan of salvation on their worlds would be to transgress against the warning and eat their forbidden fruit.

Coming to that conclusion on their own might take billions or trillions of years, or longer. Long enough for our earth to enter its creation and for Christ to be born upon it. In the case of our earth, Lucifer dramatically accelerated the fall of man by enticing our Eve to eat the forbidden fruit.

Under this line of thought, other satans would have only influenced the worlds to which the firstlings of other

Heavenly Fathers were assigned to carry out their respective atonements. Each of those atonements have eternal efficacy and infinite scope for the entire and ever increasing spirit posterity begotten by each of those Heavenly Fathers, just as the atonement of our savior does for the entire and ever increasing spirit posterity of our Heavenly Father. That is to say, those atonements extend redemption and resurrection to the spirit children who keep their first estate in their respective Heavenly Fathers' plans of salvation, just as was required in our Heavenly Father's plan.

Surely, in our pre-mortal life, we had communication with our Heavenly Grandfathers, Heavenly Uncles, Heavenly Cousins, and so forth, and knew something of their histories, just as we have association with the siblings and progenitors of our earthly parents, though we are not subject to their governing. We are subject to the governing of our own Father.

I suppose we, and Lucifer, knew of rebellions of individual sons of the morning among the spirit posterity of other Heavenly Fathers, and the atoning sacrifices of the firstborn sons of each of those Heavenly Fathers (Doctrine and Covenants 121:26-32). While we knew of such things and were aware that they constitute a pattern of events common to all newly exalted parents and their respective spirit offsprings, we had not experienced such events ourselves, until the Lucifer who was our spirit brother rebelled against our Heavenly Father.

As for the symbolism in synaptic gaps, I wonder if synaptic gaps are analogous to veils. That might also have something to do with how beings of a higher kingdom can, if they desire, visit and minister to inhabitants of lower kingdoms, including outer darkness (Christ did - He descended below all things) but are not bound by lower kingdoms. They can return to their home in their higher kingdom at will.

With the exception of righteous individuals still in their probationary period occasionally being permitted access to higher kingdoms for specific purposes, the inhabitants of lower kingdoms can never visit higher kingdoms (Doctrine and Covenants 76:111-112).

I. K. Lavanway

###

REFERENCES

"The Book of Mormon Another Testament of Jesus Christ." The Church of Jesus Christ of Latter-day Saints. 22 May 2013 <http://www.lds.org/scriptures/bofm?lang=eng>.

"The Doctrine and Covenants of The Church of Jesus Christ of Latter-day Saints." 22 May 2013 <http://www.lds.org/scriptures/dc-testament?lang=eng>.

"The Holy Bible Authorized King James Version." The Church of Jesus Christ of Latter-day Saints. 22 May 2013 <http://www.lds.org/scriptures/bible?lang=eng>.

"The Pearl of Great Price." The Church of Jesus Christ of Latter-day Saints. 22 May 2013 <http://www.lds.org/scriptures/pgp?lang=eng>.

Cohen, David. *The Adamic Language and Calendar: The True Bible Code*. Xlibris Corporation, 2009. Print.

Hilton, Lynn. *The Kolob Theorem: A Mormon's View of God's Starry Universe*, 3rd ed. Salt Lake City, 2005. PDF file.

About The Author

I am a Christian. I enjoy a youthful exuberance toward learning all I can about the Creation and Christian eschatology. I am interested in learning to understand and worthily participate in the fulfillment of prophecies and events of the upcoming days preceding and following the Second Coming of Jesus Christ.

I served as a full time missionary for two years in the Argentina Buenos Aires North Mission of The Church of Jesus Christ of Latter-day Saints, 1986 to 1988.

I am a United States citizen and a patriot. I love my country and the U.S. Constitution as envisioned by our Founding Fathers. I served on active duty in the U.S. Air Force for almost fourteen years in various non-flying operational and academic assignments.

While I never flew for the Air Force, I have always harbored a passionate interest in aviation. Since my circumstances over the past two decades have made it impractical for me to fly, I have taken up writing and discovered that I love it almost as much as I love flying airplanes.

Other books:

An Aviator At Heart (2014)

Sevenfold (2013)

Post Omerican Easter (2012)

The Modern Day Gadianton Golden Boy (2012)

Out of the Picture and Into the Picture (2012)

Platypus Boy on the Duck Farm (2012)

Duck Boy on the Platypus Farm (2012)

Earth Sink (2010)

Connect with me:

Amazon author page:
http://www.amazon.com/author/ilyan

Goodreads author page:
http://www.goodreads.com/ilyan

Smashwords author page:
https://www.smashwords.com/profile/view/ilyan

Self Publishers Showcase page:
http://selfpublishersshowcase.com/ilyan-kei-lavanway/

Facebook:
https://www.facebook.com/ConspiracyParanormal

LinkedIn:
http://www.linkedin.com/in/ilyanlavanway

Twitter:
https://twitter.com/ilyanlavanway

Blogs:
http://ebooksscifi.wordpress.com

http://conspiracyparanormal.blogspot.com

Websites:
http://ilyanlavanway.wix.com/books

http://conspiracyparanormal.com

www.ingramcontent.com/pod-product-compliance
Lightning Source LLC
Chambersburg PA
CBHW040811200526
45159CB00022B/220